New Breakthrough Science for the New Millennium

New Breakthrough Science for the New Millennium

Zero Point Energy
of Matter "Both Ends"

Raymond E. Whitson

Library of Congress Control Number: 2008903164
ISBN: Hardcover   978-1-4363-3428-0
               Softcover   978-1-4363-3427-3

This book was printed in the United States of America.

To order additional copies of this book, contact:
Xlibris Corporation
1-888-795-4274
www.Xlibris.com
Orders@Xlibris.com
47245

Contents

This book is a study manual and an introduction to a new way of thinking.

The written software in this book will answer all of the below questions, plus dozens of other unanswered questions.

* What is electricity?
* Where does electricity come from?
* What is gravity?
* How does gravity work?
* Why does a photon of light move so fast?
* Where does the photon get its continuous power?
* What causes hurricanes?
* How are hurricanes formed?
* What causes tornadoes?
* What causes a planet to rotate on its axis?
* And more.

Zero Point Energy Of Matter—"Both-Ends"

The Solar System's positive charge of 97,680 feet per second. The Uranium Atom's positive charge of 1.3 feet per second.

The greatest discovery from this study is that energy systems (when once started) will collect and perpetuate their own energy. An atom is one of these energy systems that has evolved into a close-knit family of collectors (electrons). With this knowledge, man will not have to destroy (convert) matter to collect energy.

Acquired Knowledge

This letter is to introduce the author and to show that he has been qualified to write this material. Everything that is written in this letter is true and correct to the best of our knowledge.

Raymond E. Whitson has worked at the cutting edge of technology for most of the productive years of his life. This is something that I know about, and it is a statement very few of us can claim. He is thankful that he has had this privilege, and it is his hope that this letter will encourage young men and women to step out and embrace life and to live it to the fullest in following their dreams. The last chapter in this book tells how it can be done.

During the Second World War, he was with the U.S. Marines and fought in the South Pacific. After the war ended, he joined a millwright's union and was sent to Philadelphia to help build an oil refinery. After the completion of this job, he went to Detroit and worked in the auto plants for a few years.

During this time, he had married and now had a family of six. The last two were twins. This caused him to decide to return to Knoxville, Tennessee. It was here where he applied for machinist training at Union Carbide Nuclear Plant at Oak Ridge, Tennessee. He then worked at the Y-12 plant for a number of years. During the time of a reduction in force and while talking to his wife, he decided to take a layoff slip. They both liked adventure, so they decided to move to California. They sold their home, loaded the children in the ranch wagon, and took off.

After arriving in California, they had an extended vacation with no hurry to start back to work. At that time, the companies were begging for experienced help. He could choose any manufacturing company in any town. They finally settled in San Diego, and he went to work for Solar Aircraft Company (as it was called then). One of the things that the company was working on was the X-15 Rocket. This rocket was an experiment to see if and how a rocket performs in outer space.

All of the following mentioned things in this letter he has personally helped fabricate by machining or, later on, by designing tools to do the job. These are some of the things that are on that list: the cooling shroud that covered the thruster rockets on the moon shots and some of the parts of the moon lander, the vanes and shrouds for the Pratt & Whitney turbine engines, the power stations for offshore oil rigs, the pumping stations for oilfields, and much more.

After several years with the company, he had worked his self up to the position of tool and manufacturing planner. It was a good position, and he enjoyed working for the company. The family had been in San Diego fifteen years by this time, and both he and his wife wanted to go back home and try it again, which they did. He then went back to work at the Oak Ridge National Laboratory, and worked a couple of years until the next reduction of force.

They both then decided to take a vacation to Florida. Their children were grown-up and moved away by then. While they were there in Florida, they decided to extend their stay. He went to work for Martin Marietta in Orlando, Florida, and she went to work for an elevator company. They stayed a little over a year.

Some of the things he has worked on besides the ones that have already been mentioned is the atom bomb, the H-bomb, the Persian Missile, The Sam-D Missile, and other things that he will not mention. The sequence of these happenings are very close to being right, yet there are other instances of acquired knowledge during and in between these happenings; also he does a lot of reading.

Mark Johnson

Introduction to the New Science

A thirty-five year quest by Raymond E Whitson
© Copyright 1996 under the title of
The Solar Energy Field

The complete works of this paper came through my own mind by mental concentration over a period of more than thirty-five years. This is a new and unproven science to everyone but myself. This makes me the only peer for a peer review. I say this because of our human nature. Over a period of time we come to accept as truth what we have been taught as theory. Sometimes, we were taught wrong, and it is our nature that we don't want to admit this. It is my belief that the answers to a lot of unanswered questions are in this book.

If this book does not describe the true nature of physics, then I have (in my mind) constructed a whole new universe that answers the questions of this one. Man has always been looking to the stars for answers when in fact we are already in the stars. Every thing that is out there is potentially right here with us.

Throughout this book, certain phrases will be repeated at the appropriate places. This is to keep our mind centered on this new concept.

Awareness and Perception

Knowledge is supposed to come first, and then it should be written down and passed on to a group of experimental testers before it is presented to the school board. No unproven theories should be taught or even mentioned in a conventional school because of the *brain train syndrome* (no offence intended). It is necessary to teach reading, writing, math, and also social science. It is my belief that the school system should teach students *how* to think, more so than *what* to think. They should be shown how to use their spiritual mind in conjunction with their physical brain. With this knowledge, they could accomplish most of the desires of their heart. This is explained in the last chapter of this book.

In the beginning of this study, I knew that it was necessary that I start at the beginning of *physical reality*. This means that in the beginning there had to be at least two factors of physical reality. The reason for this is because it takes at least one factor to relate to or cause reaction in the other factor. This first factor is physical energy. The second factor is physical time.

In explaining physical energy, it is energy that is moving in a circle or circuit that we will call a galaxy. This energy is timed energy, which means that it has its ends connected together forming a circle. The charge (density) is related to its time-of-revolution.

In explaining *physical time*, time is the distance of motion per density. The shorter the time-of-revolution means the smaller the system and the greater the density.

It was at this point of my thinking that I tied the galaxy to the Solar System, the Solar System to the atom, and the atom to its particles. The catalyst that causes this reduction of volume and the induction of density is time.

To shorten time is to shorten the time-of-revolution of a system in reference to its parent system (to make a smaller system inside of a larger system). Another way to shorten time is to speed up the motion of an object through the system. It has the same effect, which has to do with the total motion of the object. (This action is called kinetic energy.) All forms of motion are displacement that causes the energy or matter that is moved to combine with the parent energy that it is moving through. When it is only energy that is moving forward and combining, it is building a positive-charge on its front portion. This in turn, causes its back portion to be negative by the same amount. This positive area and negative area tend to pull toward each other. This causes the energy to wrap itself into a circle, forming a system. After reading this far, it is obvious that the third factor of physical reality is *motion.* The three factors of reality that form a system are *energy, time, and motion.* The result of this union produces charge (density). It is this density caused by motion that pulls on the parent solar field energy. The motion of the electrons pulling on the parent energy causes the systems in matter to be receptive (positive) toward its less dense (negative) parent energy. This causes the pull of gravity.

Note: There is more on these subjects farther on in the manual.

The ABCs of Energy

A

Everything in physical existence is composed of energy. This paper is an introduction to a new theory about the formation and characteristics of energy. The concepts presented here are relative and comparable with each other and answerable to the many whys concerning physics.

Energy (unlike matter) is not composed of particles. Energy is the basic substance of the universe. For our purpose, the galaxy is the first physical particle of nature, then the Solar Systems, then the atoms, and last the atomic particles. These are the four basic energy systems. A system is composed of a negative and a positive field of energy. The positive field is a circling disk of energy, and the negative field circles through this disk's center. Picture a galaxy as seen through a telescope. What we see is the outline of its positive field, which is in a disk shape with a thick center. This shape is representative of all positive fields of balanced energy systems. All of the smaller systems inside the galaxy are a reduction of space (volume) and time (time-of-revolution).

Time and space are the same thing. To shorten time is to shorten space. When time is shortened (looped into a smaller circuit), the action of energy is reversed. Instead of expanding, the energy contracts. When energy contracts, it shrinks, draws in on itself, and becomes a smaller system.

B

In this manuscript, the words *space, time,* and *energy* are synonymous. In using the word *field*, it can mean any shape of moving energy that has its ends connected and is moving in a circle (circuit). Also, for ease of understanding the concepts of energy, exchange will be illustrated using a parent-sibling relationship. The Solar System is the sibling system of the galaxy system. The atom is the sibling system of the Solar System. The atomic particles are the sibling systems of the atom. From the galaxy down, these fields will be called first, second, third, and fourth-field energy. As the fields get smaller, they also get stronger, denser and more positively charged. Electricity is third-field energy, same as the atom. A photon of light is composed of fourth-field energy. It is an atomic particle, which is a very small half-system (explanation later).

There is a statement called **the law of conservation of energy** that in effect says that it takes a given amount of energy to convert (release) the equivalent amount of work.

This is a true statement when speaking of matter. The reason for this is because matter is comprised of completed systems that have boundaries. They also have cohesive nuclei currents of energy that hold these systems together. This is why it takes the same kind of energy to release (break) these boundaries that are composed of third-field energy. The boundaries of these atoms push against and resist each other, causing some to be hard matter, some to be soft, and some to compress into a smaller area.

C

My prime subject has to do with a different kind of energy. This energy is second-field energy (solar field) or parent energy. It could also be called radio or broadcasting energy. This energy has no boundaries other than its own circling field of energy. This energy does not compress, and it does not push—it only carries and pulls. Energy is condenses (builds charge) and expands (builds motion). These two actions wrapped together form a system. The charge is the positive circling field, and the motion is the expanding negative lateral field. Without excessive resistance, these interlocked fields will continue to spin indefinitely.

All matter (third and fourth-fields) is comprised of spinning energy systems that are interlocked with rings (currents) of energy. Basic energy is the first and prime factor of material existence before it develops into matter. There are three factors that are involved in the conversion of basic energy into systems (matter). These factors are motion, time, and charge.

Motion is the action and speed of the circling field and the lateral field.

Time is the length of distance through space (time-of-revolution).

Charge is the density of the circling field relative to its portion of the radius of its field.

Third-field energy (which is also electricity) has to travel through conductive types of matter that are composed of third-field sibling systems. This electrical energy has to return to its source through third-field conductive matter (like wires or metal frames). The galaxy, which is first-field energy, and the solar field, which is second-field energy, also have to complete their circuits (loops) to hold their charges (times-of-revolution) in order to maintain their stability and perpetual motion. There is more on these subjects farther on in the manual.

D

I have read many articles on the hard-particle theory of physics, but it is my belief that all particles are composed of small half-systems that are created at the time of release from the atom. The electron pushes out these particles (half-systems) as the electron passes through the eye of the atom (spinning disk). The type of particle has to do with its charge and spin and/or volume of energy. An atom is imbalanced or breaking up when these charges are released.

Space scientists are looking for two frames of reference in space in which some form of directional motion can be induced. This problem has already been solved in the form of half-systems. One type of half-system is the smoke-ring. Before going further with this thought, let us review some common beliefs. Air molecules are thought to have powers of motion that causes tornadoes, hurricanes, and other weather problems. Air molecules only have the power to expand and/or rise up into the atmosphere. All of the weather-related phenomena are caused by second-field (parent energy) as it condenses or expands. This energy carries air molecules through all of its motions.

The two frames of reference that space scientists are looking for are enclosed in a smoke ring. A smoke ring is a half-system that is comprised of parent energy. Parent energy has a charge of 97,680 feet per second of motion. Excluding the air molecules and the smoke (gases), what is left of the smoke ring is the positive circling energy—and the interlocking negative lateral energy, the two frames of reference that scientists are looking for.

E

This manuscript tells about some tests that were made with smoke rings. In these tests, each smoke ring traveled approximately 20 feet and moved a targeted piece of paper.

This shows that these half-systems have motive power as well as kinetic energy. It also shows that the first and only impulse of energy continues to perpetuate itself until the resistance of outside forces finally wears it down.

This smoke ring system is a man-made half-system of kinetic energy inside of its parent-field (solar field), and the formula of its charge is

$$\pi \text{ x diameter x revolutions per second}$$

In this test, it was about 12.56 inches or 1.5 feet per second of motion for a 2 inches diameter smoke ring. This is a reduction of time and space of its parent solar field's energy. This smoke ring's charge is 97,680 plus 1.5 or 97681.5 feet per second of density. Its relative charge is 1.5 feet per second of motion. Its total motion, which reduces space that builds density, begins with the galaxy charge—then moves down to the solar field charge —then moves down to the smoke ring charge.

Inertia is tied to the surrounding space by the positive ring that builds charge. Impulsion is produced by the lateral, negative cohesive energy as it pulls in parent energy from one direction.

This is the way a photon of light and a smoke ring are pulled through space.

There is positive and negative energy in every system. There is also a positive and a negative charge in every system. The positive-field rings of atoms pull a negative cohesive current through their center. This action ties all of the atoms of matter together and forms a highway for electrons, photons, and parent energy to pass through. Atoms, because they are positive and *denser*, pull on the negative *less dense* solar field energy.

Note: Parent-fields are always *less dense* because the sibling systems have pulled the same energy into a smaller circle.

The Solar Energy Field

For this study, the parent field is the solar field. It is a field of circling energy that is positive and denser toward the galaxy field, which is first-field energy, and negative and less dense toward the atom fields, which is third-field energy. This solar field energy encircles the Sun. Its charge is created by the speed of its movement around the Sun at 97,680 feet per second in this portion of the Solar System.

Earth's charge is composed of multiple positive subspace systems that pull energy from the parent solar field at 32 feet per second. This is gravity. These systems of energy within, and on, Earth's surface are considered the sibling systems of the Solar System.

Kinetic Energy

Normally, when matter is stationary, the parent solar field energy is pulled into the sibling systems from all sides. When matter is moved in any direction, it is moving into the parent solar field energy, and the solar field energy is moving into the matter. This inflow of parent energy causes the positive rings of the atoms that are aligned with the movement to speed up (spin faster). This shortens their time, which causes more density (charge) on the atoms. These rings (disks) of the atom hold this charge, balance, and motion until outside resistance slows, or stops, the forward motion. As the forward motion of the matter is brought to a stop, the spins of the positive rings are slowing down and releasing their excess charge. It takes the same amount of energy to stop its forward motion as it does to move it up to that speed.

All subsystems or sibling systems within the parent field are of neutral charge. Any extra motion within the parent field builds a separate charge that is related to its own motion. This separate charge is imbalanced from its parent energy by the difference of total motion, which causes the separate charge to be stronger than the parent field charge. This stronger charge moves toward the weaker charge in whatever direction it is formed. The two most familiar examples of this phenomenon are *radio waves* and *kinetic energy*. Radio waves are composed of parent/solar field energy, and kinetic energy is formed by moving objects *through* the parent solar field. Logically, all motion should stop as soon as pushing pressure stops. In other words, a bullet should just fall out of the end of a gun. This better illustrates the grabbing and carrying power of the parent field in dealing with the internal motions (charges) of the Earth's systems (matter).

Kinetic energy is impressed energy that is collected on either the positive ring or positive disk of an atom. And whether the shape is a ring or a disk is present in a system is determined by the speed of its circling energy as it moves through the parent field. The faster the push, the shorter its time-of-revolution, which means the charge is greater. When the push on the matter is brought to a stop, the collected extra motion has been balanced by extra charge. This causes the matter to keep pulling until outside resistance balances the motional charge of the matter.

Regardless of the density or the volume of matter, it is always neutral with the parent field until outside influences like gravity or kinetic energy changes the balance of charge between the matter and the parent field. Without these influences, matter has no weight or inertial properties because energy itself has no weight. The perception of weight is only the interaction or attraction between the (relatively negative) solar field energy and the receptive (relatively positive) sibling systems. Matter has no separate entity as far as its own movement. It exists as part of (parent) second-field energy.

Medical Aspects of Second-Field Energy

I suspected that this second-field energy was the source of, or promoted, arthritic symptoms when in a low-pressure state. In order to prove the existence of this second-field energy, an experiment was set up in the following manner:

On November 19, 1995 while experimenting and searching for the cause of arthritis, I personally built a ten-foot diameter high-density expanding or condensing energy-field system. This pressure field was designed like a low-pressure system similar to what is seen on a weather channel. I had noticed that when there was a cold front associated with a low-pressure system, people with arthritis suffered more.

In my experiments, I used two mice, one for exposure and one as a control. I put the first mouse into the low-pressure field and observed the behavior of both mice afterward. The behavior differences between the two mice was dramatic, yet I was not satisfied because I had no prior knowledge of mice behavior. The behaviorisms of the mice were as follows:

Before exposure, the mice were seemingly identical in size and weight with their only difference being their color and markings. They played and cuddled for warmth. They were not difficult to catch.

After exposure, the exposed mouse would jump unexpectedly and race across its cage. It began to try to hide when I reached for it while the control mouse did not. It also began attacking the control mouse to the point of chewing until blood was drawn on its back. It was so bad that I thought the control mouse would die from its wounds. Within the two-week period of tests, the exposed mouse also appeared larger and heavier.

My next move was to place myself in the low-pressure (expanding) side of the energy system from my waist down so as not to damage my vital organs. I made a calculated guess for determining the boundaries of the field. I then placed a chair accordingly and sat on it for approximately five minutes. I either felt or imagined a slight tingling. By the second day, I could hardly get out of bed. The areas of my body that I had positioned within the field of energy, my waist and lower extremities, were swollen and I was unable to move without extreme pain. My wife had to help me get dressed and I had to buy a larger size of pants. I also wanted to sleep constantly. Then, being afraid of my condition and needing to do something about it, I forced myself to re-enter the energy field to attempt to reverse the effects. I changed the direction of the energy field from expanding to condensing, significantly reduced the density, and for a minute each day for several weeks, I positioned the lower half of my body in the condensing field of energy. The pain and swelling gradually went away.

During these weeks of healing, I had a long plane trip to make, and I was dreading to make the trip. The reason for the dread was because of how painful it was for me to sit down. To my surprise, as soon as the plane got up speed, the pain left me, and I felt great. I then realized that it was the kinetic energy (parent energy) flowing through my body that relieved the pain. This is the same energy that causes jetlag in airline passengers. This energy causes the sibling systems in our body to draw in and convert more parent energy to third-field density. This is the excess energy that throws a person through the windshield of a car during a crash.

As soon as we landed, the pain came back but not as strong. Since I have never had arthritis, the reaction of my muscles to the expanding side of the energy system (in my opinion) proved a significant finding about the relationship between the expanding side of a system and arthritis.

Several months later I immersed my dog (Buddy) in the condensing side of the system for ten minutes at full density. It gave him a lot more energy than I realized at the time. The next morning, he came into the house with his tongue hanging out. He had these sticky burrs all through his fur. He had mud all over him, and he was sloppy wet. I believe that he romped through the woods and fields all night long. The other strange thing that he did was flop down in front of the air conditioning furnace and sleep all day. He slept so soundly that I could hardly wake him. I had to speak to him and nudge him with my toe, and then he would open one eye and wag his stub tail a couple of times, then fall back into a deep sleep. When he finally woke up, he was still a happy healthy dog. After a few weeks, I saw that his experience had not harmed him, so I immersed myself in the same field for 20 seconds. It seemed to give me more energy and make my breathing easier. But after some shopping and some walking in the park, I got real sleepy like I had jetlag. After sleeping soundly for about an hour, I felt great.

Explaining Experiments

The second-field, parent-field, and solar-field are just different names for the same field. When the energy systems within a body (matter) are brought into a parent field that has more positive charge (is denser) such as the energy field I created, the electrons slow down and drop into lower orbits, yet maintain their same time-of-revolution. They need less motion to collect their usual amounts of energy (explanation farther on).

Because I was in the expanding side of the energy system when I made my first test, it drew third-field energy from my body's atoms. When I stepped out of the denser field into normal second-field energy, the electrons became locked in a lower orbit. This inability of the electrons to quickly readjust is what causes muscle atoms to draw toward each other. This energy drain has little effect on most matter, but muscle atoms have a large range of contraction and expansion. (Note: Positive energy shrinks when accelerated. Negative energy expands.) The action of expanded second-field energy (negative energy) on muscles is to draw third-field energy from the atoms that it touches. This prevents the atoms from functioning properly. This causes the arthritic symptoms of pain and swelling.

These experiments with the expanding low-density side of the system and the experiments with the condensing high-density side of the system had a twofold purpose. The first purpose was testing for the cause of arthritis. The other purpose was to prove the existence of second-field energy (solar field). This second-field energy is the energy that causes tornadoes and hurricanes, pulls sailboats, and lifts airplanes. It also carries radio waves, and it carries gas atoms toward the earth.

The reason this energy is so hard to understand is that it has only two handles by which we can grasp it. The one handle is air molecules (gases) and the other handle is radio waves. All other matter, both solids and liquids, are connected with rivers of cohesive current that carry second-field energy through them. Gases are separate systems complete within themselves. In their natural state, they do not join—they bump against each other. Their charge is so near in strength to the parent field charge (solar field) that they draw very little energy. For example, wind is composed of moving gases. The parent field does not go through gas as it does connected matter. Instead it carries the gas molecules.

When the parent field energy moves, air molecules (or gases) move with it.

When gases are moved, they displace parent-field energy.

Gas atoms are free to align with the fall of gravity. Their expanding side (intake side) faces up, which allows them to climb or rise up whenever heat is added to them. They also are subjected to a constant pull toward the earth by the flow of the parent field energy (gravity) as it flows into the earth. The greater the difference of density (charge) between the parent field and the sibling fields means the greater the attraction between them. This is why gases have very little attraction toward the negative pull of the parent energy. As matter is pulling itself toward the Earth (as explained later) or pressed against the Earth, then all of the parent field energy that is between it and the center of the Earth is less dense because this portion of parent energy is being converted to matter. (It is more negative.) This also causes the atoms of matter that are deeper in the Earth to be deficient in parent energy. This shortage of parent energy causes these atoms to expand (reach for more parent energy). The product of this action is called hot lava. Radio waves also move through the parent energy field without displacing the parent energy. These waves, of all kinds, are composed of charges (densities). This action is explained in the Deficient Parent Energy section.

Understanding an Atom

In understanding an atom, think of a smoke ring. As the smoke ring moves away from you, the smoke on your side of the ring moves toward the center then through the ring to curve back around the outside perimeter toward you in completing its circuit. Now, think of a smaller smoke ring going through the center of the larger ring in the same direction. Since energy will not move against itself, the outside perimeter energy of the smaller ring has to go all the way around the outside perimeter of the larger ring to complete its circuit. This causes expansion and high speed of motion, both of which draws parent energy. This smaller ring represents the electron in an atom, and the larger ring is the positive-field ring (disk) of an atom. Now think of these electrons as small rings that are threaded onto a large slender key ring. These electrons are equally spaced on the rings of negative lateral energy that rotate through the center of the large positive-field ring. When a system gains one ring and one electron, it becomes a Hydrogen atom and the beginning of third-field energy.

In the denser atoms, the electrons move between rings freely as the charge demands.

All electrons cycle through the center of the large positive-field ring, always moving in the same direction and always cycling in the same period of time, relative to their identity (element). When the lateral negative rings are expanded (lengthened) as in a boundary gap, the electrons have to speed up to hold their time of cycle constant, which also causes them to draws more parent energy. This extra motion of the electrons happens in the atoms that are pulling toward the deficient boundary gap side of the matter. This means that when matter is falling toward the Earth, what it is really doing is pulling itself toward the Earth. The boundary gap is the area below the bottom of the falling objects. This boundary gap extends to the center of the Earth.

Third-Field Energy

When the electron orbit rings that pass through the center of an atom are broken, the density (charge) of an atom is released (expanded). The released energy goes back into the same amount or volume of parent-field energy that it was taken from. As the expanding energy reaches the same density as its parent-field, it will stop moving on its own and continue moving with its parent-field. At this point, it has turned back into and become second-field energy again.

There are three ways that the charge of an atom is released and turned from third-field energy back into second-field energy again.

1. Natural decay (rotting)
2. Burning (as burning gasoline)
3. Exploding (as setting off an explosion)

Fourth-Field Energy

A photon of light is a half-system that is composed of fourth-field energy. It is denser (has more positive charge) than an atom. This causes its negative energy to expand and speed up in its effort to draw enough energy from the second field (its grandparent-field) to supply its fourth-field charge. This condition of balance is never met.

- A photon of light is in a state of continuous explosion in one direction.
- It can exist only as long as it is moving through energy.
- If the energy that it is moving through is denser, the photon of light moves slower.
- If the energy is less dense, the photon moves faster.
- A photon of light, like all particles, is only a half-system.

When half-systems are brought to a stop, they disintegrate, and their energy is absorbed by adjacent matter. The difference of charge between a photon of light that is composed of fourth-field energy and the solar field (which is second-field energy) gives the photon of light a speed of 186,300 miles per second through the solar field's density, which is its slowest speed. Its speed is greater in the galaxy and even greater in-between galaxies. A photon of light coming into the Solar System slows down because of the density of the solar field. This slowing down of the photon causes a Doppler effect on the spectrograph that makes the galaxies seem to be receding. This receding effect combined with the belief that photons of light move at the same speed throughout space is what started the big bang concept.

Full Systems

A full system consists of a circling disk-shaped ring of positive energy with a hole through the center. It also has a negative energy field that moves in from the outside perimeter and flows toward and through the center hole on one side of the disk. As this energy leaves the center hole on the other side of the positive disk, it expands across the complete face of the disk and speeds up in its trip back to the outside perimeter where it completes its circuit. These actions (expansion and motion) collect parent energy on the front side of the disk, and then as this collected energy goes around the outside perimeter of the disk and starts back toward the center hole, it condenses (builds charge). This looping action loops the collected energy into a circuit that resets its time and its charge (density). It is this action that changes the second field/parent energy into third-field energy.

Half-Systems

- Atomic particles are created in an atom.
- Their parent-field is the atom, which is third-field energy.
- Atomic particles are therefore fourth-field energy.

As long as atomic particles are in the atom, they are still full systems. As soon as they are released out of the atom, their parent-field is now the solar field (second-field energy). This causes their energies to expand and speed up, trying to draw in enough parent energy to sustain their density. All systems draw in parent energy from one side. This causes the half-systems to fly through the solar field (grandparent energy) at a terrific speed, trying to draw enough negative motional energy to balance the positive density of their positive fields. They have now become unbalanced half-systems.

Some half-systems are created by outside pressure. Kinetic energy, which is caused by moving physical objects through the parent field energy, is one example of this action.

Another example of this action is the smoke ring, which is a half-system that is created in this way. Every system has a separate and greater charge than its parent-field, which in this case is the solar field. Smoke (air, gases), when rolled off the lips of a smoker, or forced from the end of a round tube, causes a ring of parent-field energy to form into a half-system. Because this smoke ring half-system has a separate charge from its parent field (solar field), the gases of the smoke ring are also sibling systems of the smoke ring that is now also a parent field.

In an experiment I made to further this study, I made a smoke cannon. I used a piece of 4" PVC pipe about 2 feet long. In one end, I placed a plunger with a handle that stuck out through an end cap. In the other end, I fastened an inside lip formed out of plaster. By hitting the end of the handle with a hammer, I could make half-systems travel about 20 feet and strike a sheet of paper. To make the smoke (a smoke generator), I placed a soldering gun in a can of oily rags and made a tube-fitting for the can lid and PVC pipe. I then piped the smoke up into the side of the PVC pipe (cannon). After I had the cannon aligned with the sheet of paper, I did not need the smoke anymore to hit the paper.

Another test I made with the cannon was to send smoke rings through a curved pipe. I used a 3" diameter electrical conduit elbow that had a smooth curve. I attached the curved 90-degree elbow to the cannon. When I struck the plunger, I expected a puff of smoke to come out of the end of the curved pipe. Instead what came out was a perfect smoke ring, time after time. This is still a mystery to me. The fact that the rings came out straight and true shows me that systems and half-systems will align up with their forward movement. This alignment of sibling systems is explained in a previous chapter. It is called kinetic energy.

Another discovery I made about half-systems is that they don't pull parent-energy into themselves the way an atom or a full-system does, instead they pull themselves into and through the parent-energy. This allows the speed of their forward motion to collect extra negative parent energy. This extra motion keeps the positive disk balanced. This is the reason that a photon of light moves so fast. The more negative (less dense) the parent-field that it is moving through, the faster the photon moves through the field.

Smoke Ring Cannon

Spacecraft

Whenever man builds a true spacecraft, this is how it will have to work: he will fly through space in a man-made energy system (a man-made photon). When the negative lateral energy and the positive-field ring energy are balanced out, the craft will remain stationary. When the motion of the craft's negative field exceeds the motion of falling objects, then all of the feet per second greater than that is lifting speed. Keep in mind that matter has no "weight." It has only an attraction toward the negative portion of its parent-field's energy. Its parent-field is now the energy that surrounds the craft. It has become an independent system of its own. All of the craft and everything that is inside of the craft becomes as one unit—free from gravity, free from inertia, free from the planet, and even free from the Solar System. While it is in the Solar System, its top speed will be somewhere near the speed of light. After leaving the Solar System and moving into the Galaxy System, its speed will jump to somewhere near the *speed of light squared*. This happens because of the less density of the galaxy field and because it will be doing the same thing that a photon of light does. A photon of light *squares its speed of motion* when going out into a *less dense system*, and it drops back to its *square root of motion* when entering a *denser system*.

When matter moves *through* parent-energy, the parent energy moves *through* the matter. This action is what causes kinetic inertia (also jet lag). Objects that are carried *inside of a system* have no inertial reaction to the change of direction *of the system*. When a smoke ring system goes around a corner, it carries the gases (smaller systems) around the corner with it (as explained in the section on half-systems). This means that a photon-craft (a craft that is shaped *like* and surrounded *by* a photon) would have no inertial reactions when changing directions. We can also presume the following statement: In daylight and when this type of craft is making fast motional maneuvers or high speeds, the photon image of the craft cannot escape through its outside field of energy. This would cause it to become invisible to the naked eye. This distortion would be similar to the distortion that we see when looking at something through heat waves, only more so.

Another example of this action is our Solar System. It carries the planets around the system in a belt of like-charge. This is demonstrated by the action of the asteroid belt, which shows that these pieces of matter have the same density charge that is related to that radius of the Solar System. This belt of circling energy is more positive on the inside *toward the center of the system* than it is on the outside farther away from the center. This gives the pieces of matter a charge that is related to that particular belt of energy that is carrying them. Remember that matter has no weight. It only seeks a positive and negative balance of its internal systems. At one time these pieces of matter were joined together as a planet within this same belt of energy.

See the section on Density, Time, and Motion.

Deficient Parent Energy

The shortage of charge (density) in the parent solar field causes a shortage of charge in the atoms of matter. This shortage of charge means excess space or distance, which means larger orbits of the electrons. Larger orbits of the electrons mean more distance to complete their set time of timed revolution.

- More distance per time means more negative motion.
- More negative motion pulls in more positive charge, which rebalances this shortage.

From this, we now have a gob of matter that is stable in an unstable condition. This is the condition of our Sun. It is also the condition of hot lava. The difference between the two is that the lava rehardens when it reaches the surface of the Earth. Here on the surface, it gets more parent-field energy (falling gravity), which brings it back to its original charge.

The parent-field energy encircles the Sun. This leaves the Sun in outer space where it is fed by the galaxy-field energy, grandparent-field, which is first-field energy. This energy will only hold matter in an expanded or gaseous state. The exception to this is when matter is moving *through* this energy with a certain velocity, which builds motional charge. The shortage of charge (density) of the galaxy energy causes expansion and motion of the sibling systems of the Sun.

The lateral energy that circles through the center of the Solar System adds enough charge (density) to the matter (sibling systems of the Sun) to keep them from breaking up (exploding). The excess volume of energy is converted to photons of light (i.e. the energy flowing through the Sun causes it to burn like a lightbulb). The Sun is not in our parent-field. The Sun is surrounded and entrapped by our parent-field. The Sun is what sustains our planet life as we know it. The Sun has no gravitational pull on our planets because matter does not pull on matter. Matter only pulls parent energy toward itself. Parent energy is the energy that is touching the matter. The parent energy of the Sun is the galaxy energy.

Note : *Magnetism is a negative cohesive third-field energy that is driven out of normal bounds by over charged positive fields of the atoms.*

Space-Time

Solar field and Atom

Time, motion, and charge are all attributes of energy.

- Moving time (timed revolution of solar field energy) travels in a circle.
- Moving time has a set circumference of 360 degrees of motion to complete its density charge and to set its time frame. It then becomes a system. After passing the cycle point this energy will continue moving at the same speed until outside forces interfere. Any more change in motion in any direction by any part of this energy breaks this balance. The part of this energy including matter (atoms) that is moved adds more motion (distance) to this portion of the moving energy. This added motion, adds on to the solar motion of 97,680 feet per second. This shortens the time (time-of-motion per distance) for this portion of the energy. This builds charge. We call this charge

kinetic energy. Motion causes more positive charge on the atom. This, in turn, makes the negative parent energy relatively more negative by the same amount. This keeps the two systems pulling toward each other. One is an expanded negative charge, and the other is a positive pulling charge.

- More distance per time means less time per distance.
- Less time means less space.
- Time and space are the same thing.
- To shorten time is to shorten space.

When space is shortened, it shrinks, draws in, and becomes denser. The amount of time (space) that is shortened make the atom's positive fields denser.

As this parent energy comes into the Earth, it is converted into smaller time capsules (systems and subsystems). The distance that is added and the time that is shortened are equally proportionate to each other: to make half-systems such as low-pressure systems (caused by energy going into the Earth), or to make high-pressure systems (caused by energy coming out of the Earth), or to make smaller sibling systems.

When energy moves in a forward direction, it also causes energy to circle counterclockwise around the forward moving energy. This is also the left-handed rule of current and magnetism. Pointing your left thumb toward the face of a clock and curling your fingers clarifies the directions of these motions.

Short Version of Space-Time

The motion of energy or the motion of matter causes it to separate itself from its parent energy and take on an added separate charge. The added separate charge is only justified by the added separate motion. These two actions of energy—motion and charge—working together are the building blocks of the physical universe. These actions are visibly seen and mentally witnessed as the motion of satellites and the motion of radio waves. There is also another very important reason for these two actions of energy. When these two actions are interlocked (linked together) they form stationary systems and stationary atoms relative to their local surroundings. (Half-systems are explained elsewhere.)

Looking Down

The galaxy-field energy systems are the least dense of all the energy systems. The solar-field energy systems are denser than galaxy-field energy systems. The atomic-field energy systems are denser than the solar-field energy systems. The particle-energy systems are denser than the atomic-field energy systems. As the density of energy becomes denser, the *time*-of-revolution gets smaller, and the occupied space gets smaller. The end product must be somewhere near the density of the man-made particles (half-systems) caused by cyclotron bombardment.

Looking Up

The first product of controlled revolution and time (so far as we knew before the Hubble telescope) were the galaxies. Now we see larger systems that are not as dense and have a lesser charge than galaxies. These huge systems act as lenses that magnify and draw in the light from more distant galaxies. These systems that pull light photons through them in one direction will also push or prevent light from moving through them in the opposite direction because of the lateral negative energy that moves through the center of all systems. These huge systems could also be called black holes when looking at the back side of the system.

The Flip Side of Energy— Magnetism

Energy has the same power to contract as it has to expand, burn, or explode. This negative side of energy is the most important and least known side. It is the creative side of energy versus the disintegrative side. This is the energy that holds matter together and causes the planets to be spherical. It also causes the pull of gravity and sustains the life of all plants and animals. The most familiar and the most used contraction of energy is magnetism. Magnetism comprises the negative portion of third-field energy and density (charge) comprises the positive portion of third-field energy.

All negative lines of energy (currents) in matter are composed of what we call magnetism. These lines of energy match the charge (density) of their specific elements. They also match a blend of elements with the same charge. This energy should be called third-field negative energy, and electricity should be called third-field positive energy. As the negative lines of energy pull themselves into a circle, they become timed (time-of-revolution). This action has to do with the distance of travel around the circuit and the speed of the moving energy. The motion of third-field energy pulls in parent (solar field) energy. I call this action second-field energy that is being converted to third-field energy. This same action is what happens in a conventional generator when electricity is collected.

Electricity is second-field energy (parent energy) that has been pulled in and retimed to third-field density, retimed from 97,680 feet per second of density to less than 312.5 feet per second of density. This is explained in a later chapter under "Energy Potential."

Magnetism is negative third-field energy and has no explosive ability of its own. Instead, when broken up, it implodes and is absorbed by the expanding (exploding) positive field. An atomic blast is caused by the fusion of the two fields, the positive field and the negative field. It takes the two fields of energy to have an explosion. One field collects, stores, and holds the energy of the other field in bondage until its own field is broken.

A lightning strike is caused by the negative energy in the earth jumping to the clouds, as proven by scientists and technicians in the field. When this happens, the negative energy breaks (releases) a bond of positive energy from the clouds that follows the strike back down to the earth as they fuse together, causing an *implosive* action. This action shows that energy has the same power to contract as it has to expand, to implode as to explode, and to create as to disintegrate.

The contraction of second-field energy (parent energy) is what causes all of our weather. It also forms and sustains the sibling systems of the earth. It forms radio waves, and it forms boundary gaps (gravity) of great pulling strength.

- Space and energy is the same substance.
- Time is the controller of charge that separates the different densities of energy.
- When time is shortened, the action of energy is reversed; instead of expanding, the energy contracts.
- When energy contracts, it shrinks and becomes denser (has more charge).
- Every time a system is formed, it is a new timed capsule. It is the basic unit of existence and material life as we know it.

When this progressive reduction of the size of the systems reaches the atom stage, which is the third field, these systems start connecting together with cohesive nuclei currents of energy. These systems join together according to their charges and combination of charges.

The Attraction between Matter and Energy

When a portion of the parent energy is moved, that portion becomes denser (more charge) as it draws in (takes from) the surrounding parent energy. This action forms a gap or boundary of negative space (energy) that causes adjacent matter to move into the gap.

An airplane wing creates the above boundary gap of negative space that lifts the plane's wing. Air molecules moving over the top of the wing move two times the thickness of the planes wing faster than the air that moves under the wing. Actually, the air on top of the wing is the only air that moves. The air on top moves up and lets the wing slide through and then moves back down. When air (gasses) are moved, they carry (displace) parent-field energy. These moving air molecules create a stream of (a portion of) the parent energy. Because this portion of parent energy is moving faster than the rest of its field, it forms a separate charge, a greater density, which pulls in the adjacent parent energy. This action forms the boundary of negative space that lifts the plane's wings.

Rocket Thrust

There is a statement that says *"For every action there is an equal, and opposite, reaction."* This statement refers to third-field energy and third-field systems. The parent energy, which is second-field energy, works the same way, only in the opposite direction. When parent energy moves, it shrinks and pulls energy toward itself.

The motivation of a rocket is caused by the displacement of parent energy. The motion of the gases carries a stream of parent energy through the thrust chamber. This action pulls a boundary gap of negative space in front of the thrust chamber and in front of the rocket. This boundary gap pulls the rocket into the gap.

The gravity of Earth is another example of the contraction of energy as it pulls objects toward the Earth. All of the electrons in the Earth are moving and drawing parent energy into the Earth. This causes a negative boundary gap around the Earth that is less dense. Objects that are denser and have more charge (more matter) are drawn toward the negative less-dense boundary that extends to the center of the Earth.

Rotation of the Planets

As the parent energy spins around the Sun, the energy that is closer to the Sun spins faster than the energy that is farther from the sun. This is a progressive action that extends from the outer planets to the inner planets. It is this action that causes the planets to rotate on their axes. The parent energy that is touching the near side of a planet closer to the Sun is moving faster than the energy that is touching the far side of the planet farther from the Sun. This causes a twist or roll to the planets, like rolling a marble between your fingers. The larger the planet means the greater the spin of the planet because of the greater difference in speed between the two sides.

This faster moving energy on the near side of our planet has a slight drag on the surface of the Earth. This is what causes our weather pattern to move in an easterly direction. During the hurricane season, when our planet is angled in the right position to the sun and the ecliptic plane, this same energy interacts with the curvature of the Earth and its rotation. This interaction can create a counterclockwise spin of a large volume of air that causes the start of a hurricane.

A cyclone can happen six months earlier, or six months later, on the other side of the Solar System and on the other side of the equator. The energies are then moving in the opposite direction. This is because of the tilt of the Earth and its tendency to always lean in the same direction like a gyroscope. The only difference between a cyclone and a hurricane is the direction of their spin.

Solar field energy (second-field energy) causes our atmosphere (third-field energy) to move around the world in an easterly direction as the world turns. This circling third-field positive energy causes third-field negative energy (magnetism) to circle through the center of the Earth forming our magnetic field.

Tornadoes

A tornado is a stack of smaller half-systems that are caused by a release of parent energy through a hole in the clouds (just like flushing a toilet). Some clouds have more resistance to the fall of parent energy because of the density of the separated molecules. This causes the clouds to be pulled closer to the Earth until the pressure is released. All streams of forward moving energy (especially toward the ground) cause counterclockwise energy to circle around the forward-moving energy. This action of energy is what builds all sibling systems from the hurricane down to the smallest particles. This left-handed motion of energy causes the lateral energy of a tornado to flow down through the tornado along the inside wall into the ground and then curl out and back up toward the top, lifting loose debris as it rises. The strongest portion of this lateral energy is close against the inside wall coming down and the outside of the wall going back up, where it completes its cycle. This cycling lateral energy tries to complete its cycle in the same period of time as the circling energy of the funnel. This means that the longer the funnel, the faster the motion of the lateral energy. This extra motion of the lateral energy is transmitting to the circling energy. This causes a shortage of time-of-revolution (space) that shrinks the systems in the tornado as the tornado reaches toward the ground.

This action forms the common funnel shape that we see in the longer stretching tornados. This fast motion of second-field (parent energy) forms a boundary gap of negative space that jerks objects of matter toward it and into the turbulent motions. This pulling action of the boundary gap is the same action that happens on the top of an airplane wing that lifts heavy airliners into the air.

Note: Forget the wind and water of a tornado or a hurricane. They are only carried objects. They have nothing to do with the power of energy systems.

Falling Straight Up

To build a spacecraft, it is necessary to construct an energy system (like a tornado) into a ring by connecting both of its ends together. When these ends are connected, it then becomes a completed system. The circling energy then becomes the lateral energy, and the lateral energy becomes the circling energy. This new lateral energy condenses on the bottom side of the system and expands on the top side of the system. As sibling energy expands, it exchanges its density for motion and expansion. This extra motion of the sibling energy shortens its time (space), which causes it to draw in more parent energy. This leaves a partial void (a boundary gap of negative space) over the top of the craft that pulls the craft up, which in turn raises the boundary gap as it goes. Since this boundary gap is greater (has a more negative space) than Earth's gravity and since the craft is enclosed in the system, it will cause the craft to fall straight up away from the Earth at the speed of its lateral energy minus the speed of 32 feet per second, which is the fall of gravity.

We must also keep in mind that air has about five psi pressure at jet cruise altitude. This means that at best, we should only get five psi of lift on top of the airplane wing at that altitude. What I'm trying to get across is that gravity can be pulling from any direction. The pull that lifts an airplane's wings is the same as the pull that tries to pull the plane down. The only difference is that the gravity on top of the wing has the greater pull, and it is only an inch or two high (thick). The following is an explanation:

When air molecules are moved, parent energy is moved with it. When parent energy is moved, it condenses (shrinks). When parent energy shrinks, it pulls in the surrounding parent energy, causing a boundary gap of energy that is partially depleted of energy. This means it is more negative and less dense. This causes the wings of the plane to be relatively more positive and denser by the same amount. This causes the positive wings to pull toward the negative boundary gap, lifting the plane as it goes. This also explains why we can continue to make heavier aircraft and they fly. The greater the mass or density means the greater the pull on the boundary gap. Matter falls toward the strongest boundary gap in whichever direction that boundary gap happens to be.

The motion of the electrons in the Earth pulls parent energy into the Earth. This forms a boundary gap of negative space (energy) in and around the Earth that causes objects to pull toward this gap.

Energy Potential

The uranium system (atom) will be used as an example with its atomic number of 92 electrons and its mass of 238.07 (atomic weight). When figuring the volume of potential energy, it is calculated as $E = MC^2$. This means that each of the 92 electrons in the uranium atom has pulled in the parent energy (second-field energy) and condenced it down to 2.8% of its volume of space. This energy when released, moves back to its former charge and motion of 97,680 feet per second as it moves around our solar energy field.

List of Abbreviations

An—Atomic number (Uranium—92 electrons).

M—Mass (atomic weight) U-238

Sf—Solar field—97,680 feet per second of motion charge.

D—Density FPS (feet per second).

By using the factor of charge (density) instead of the speed of light through that charge, the formula for potential energy reads as

$$E = MD^2$$

The formula to find the density of an atom reads as

$$D = S_f / M$$

This is the density of matter before it is expanded. For example, the square root of the solar field at 97,680 feet per second of motion is 312.5 feet per second. This divided by 238.07 (U-238 mass) equals 1.313 feet per second of motion. Remember, the smaller the system means the greater the density, and the slower the motion of relative energy in that system. The uranium system (atom) has a charge of 1.3 foot per second of circular motion on its positive field.

- Density before explosion: $D = S_f / M = 1.313$ FPS of circular motion
- Density after explosion: $E = M D^2 = 97,680$ FPS or 18.5 miles per second as it moves around the Solar System

This is the speed (charge density) of the parent field energy in our portion of the Solar System. *This formula is not to show accurate measurements, only accurate relationship.*

The released energy of all elements moves back to the speed of the parent solar energy of 97,680 FPS as it moves around the Solar System. The amount of released energy for each element is calculated by the number of units of pulling power (mass) drawing from the solar energy field. The hydrogen atom is the first element to represent this calculation. It has one electron that condenses the solar field down to 312.5 feet per second of density. This is basic third-field energy. All of the other atoms build onto the hydrogen atom. Each electron that is added shortens the time (space) of its now parent-field. As these electrons are added to the hydrogen atom, they pull its field into a smaller circle, which changes its time-of-revolution and its density. It also gets another name. It becomes another element. As the electrons are added, they reduce the volume of space of the previous field by its square-root. This builds charge (density). In the uranium atom, this happens 92 times.

Internal Functions of a System and Their Common Names

A system is a gateway between two different fields of charge. One field is the positive sibling-field (atom), and the other field is the positive parent-field (solar field). The positive sibling-field has the greater charge, which we can call the square root of the parent field charge. *(Referring to the change between second and third-field energy.)* The positive field of a system is a semiflat circling ring (disk) of energy that progressively thickens toward the center hole. It also has a reduced time-of-revolution for each incremental portion of its radius as its radius extends out toward the diameter. (More space means more time.) Time and space are the same thing.

When motion is physically added to matter (as being pushed), the positive rings of the atoms that are in line with the movement draw (collect) extra second-field energy. This energy that passes through the center of the positive rings causes them to spin faster, which shortens their time-of-revolution and builds positive charge. This extra charge is called kinetic energy, and it balances the motional negative energy that has been added to the matter. When heat, which is extra third-field energy, is added or taken away from the atom, the whole system expands and speeds up, or contracts and slows down. This keeps the time-of-revolution the same and the charge the same. *It is still the same element.* The only thing that is added or taken away from the atom is the volume of energy and the motion of energy. The larger volume has to move farther and faster to complete its circuit time (time-of-revolution).

Heat is excessive third-field energy per positive charge. For lack of positive charge (density), the atom (system) expands and lengthens the distance of its revolution (per time) and generates a high-speed motional charge to balance itself. As the electrons of these atoms pass through the center of the positive-field ring (disk), they push out this excess energy in the form of infrared photons. This high-speed excessive third-field energy is what pulls atoms apart when matter is cooking or burning.

Systems in Transport

Keep in mind that the words *sibling systems* or *atoms* mean the same thing, and that these words are interchangeable with the word *matter*. By using the charge of the Solar System at 97,680 feet per second of density and by finding the density of an atom, we can deduce the transporting potential between the two fields. The charge of an atom grows with the mass of an atom. This causes different atoms to have different strengths of attraction toward the parent field, according to their number of electrons. More electrons mean more pull on the parent energy. (The conventional way of saying "it weighs more.") The greater the difference of density between the solar field and the atom field means the greater the attraction between the two fields. A sibling system draws these two fields together, forming an impressed tension between them that is released instantly by an explosion, or gradually by burning, or as with half-systems, by neutralizing their charges with in-line motion. This in-line motion is a controlled continuous exchange of charge for motion and motion for charge.

Everything in physical existence is explained with this principal of energy exchange, including the inertia and kinetic energy of moving objects (as expounded in the Kinetic Energy section). Even though these are both positive fields—solar field and atom field—they are separated by at least a square root of density. It is this difference of charge that causes this potential of attraction between them. The negative (lateral) field of a sibling system surrounds and covers the positive field of the sibling system and cancels out this attraction. This lateral field forms a motional negative charge that balances the two positive fields by forming a balance of motion-to-charge ratio in the atom. This allows the sibling field to have a greater positive charge than its parent-field and yet be neutral with its parent-field.

The Solar Field and the Motion of the Planets

The two things that change the neutrality between the parent system and the sibling systems are kinetic energy as in moving objects or a boundary gap (like Earth's gravity). Here, the parent energy is pulled toward the Earth, causing an expanded negative charge of the parent field energy. As this portion of deficient parent energy flows through matter, it leaves a boundary gap of negative space on the side of the matter where the energy exits. This causes the systems to transport toward the negative space, but only to the point of filling the space. This is why all objects transport toward the Earth at the same speed (minus air resistance). They transport just fast enough to keep their own boundary gap closed. The greater the density or the volume of mass means the greater the pull. This is the same action that pulls the planets around the Sun. Gravity and parent solar-field energy are the same energy, making the same energy exchanges, and making them in the same place. Part of this energy is moving toward the earth and carrying objects of matter with it. This energy and the rest of our portion of the solar field energy, including the Earth, is moving at 97,680 feet per second around the Solar System.

- Negative space transporting is caused by a condition of the parent field that is called a boundary gap, or with photons of light is called a system gap.
- Inertial or kinetic transporting is caused by a state within sibling systems. It is a directional charge on the sibling systems that pull their own boundary gap.
- Gravity is a single directional pull that causes the sibling systems (matter) to chase the parent field energy as the parent energy moves toward the Earth.
- The sibling systems of the planets also do the same thing as they chase the parent energy around the Solar System.
- Sibling systems that are deep within the Earth have a multidirectional pull that I call reaching, which is also caused by a shortage of parent energy. This shortage causes the atoms to expand into molten lava. (This action is explained in the Deficient Parent Energy section.)

Density, Time, and Motion

When space (energy) is wrapped in a circle (as in a system), the *time* of that system is the time of each incremental portion of its radius of revolution as the energy extends out from the center of the system. Each portion has its own time-of-revolution, density, and charge. Time is an actual physical controller of energy. It controls by its *time*-of-revolution and its *radius* of revolution (which relates to distance of motion). This, in turn, also controls the density of the energy in each portion of the radius of the system from the center all the way out to the rim. Therefore, each planet should have its recorded time of motion changed to match its true time instead of using Earth time. The sibling systems in each planet have their own charges relative to where their planet is located in the solar field. The third planet from the Sun (Earth) seems to have the right time-charge (density) for supporting life. Farther out in our Solar System, where the *time* of the orbiting solar field energy is slower, the atoms of matter are more extended and are mostly lava and gases.

Note: *All large bodies of matter (planets) have only molten lava beneath their mantle surface crust. This is because of the shortage of parent density as the planets grow and move out from the center of the system. (This is explained later, under the heading of "Growth Of Matter.")*

Jupiter is the first planet to reflect this lack of density. The notable, large red circle is a hole or window created when either the crust was blown out, or more likely, just collapsed. The next three planets—Saturn, Uranus, and Neptune—have no solid matter under their gases. Only their moons have solid surfaces, and this is because of their size and the extra motion of their orbits around their planet, which builds kinetic energy (charge). Pluto, which is called the farthest planet from the Sun, is not a planet, but a satellite of the Solar System. A planet grows only inside of a star system. Pluto gets most of its density from its plunge through the outer fringes of the solar field. It also has a constant charge of more than 10,000 miles per hour of motion while outside of the Solar System. This outside energy is less dense, which means less charge per foot of motion. Pluto's density charge is gradually reduced as it travels outside the solar energy field. This causes it to expand during this time.

In the first part of this chapter, it explains how the charge of sibling systems (matter) is related to the diameter of orbit of its planet in the Solar System. This same law also causes the planets and all parts of them to stay in their own orbits. With this thought in mind, consider the asteroid belt. With all of these pieces of matter in the same orbit and the fact that planets grow in a round shape, it is easy to theorize that these pieces of matter were at one time a planet. If this is so, it is also easy to think that Pluto entered our Solar System and collided with a planet in the orbit of the now asteroid belt. This could be what caused Pluto to take up a tilted orbit and what caused the belt of asteroids. If parts of this planet struck the Earth and/or Jupiter, it could have been what caused the deaths of the dinosaurs and/or the hole in Jupiter.

Note: Most of these statements (until proven) are based on the principles of this new theory. The most visible proof of this theory is the fact that the Sun, which is outside of the solar field and which is stationary in relation to the solar field, is in a state of balanced expansion as explained in a previous chapter. Physical bodies can hold all of their density in a straight line of motion when outside of their system if they move fast enough. This is because of the inductive action of sibling systems in matter. The motion of energy shortens the time of its energy and builds density and does it automatically. This is demonstrated in the speed of a lead bullet as the bullet penetrates an object that is harder than the lead of the bullet.

Warp Speed

The density of a sibling system is equal to at least the square root of the density of its parent-field. When a half-system (like a photon of light) with its square root charge is released into its grandparent-field's normal charge, it causes the half-system to square the speed of its motion every time it passes into a more expanded parent-field. This would square its speed at least twice by the time it left the galaxy. This is at least 345,960,000 miles per second. Now think in terms of space energy: since there is no moving time outside of a galaxy, straight line motional charge is all that is left. The motion of energy probably starts at the maximum end, which is instantaneous, and then gradually slows down as time and charge are added. We are living close to the zero end of motion (like the charge of a uranium atom at 1.3 feet per second of motion). This means that the motion of a photon of light speeds up as it approaches zero charge and zero time-of-motion, which means maximum motion. Between the galaxies, motional charge replaces density charge. Motional charge starts in maximum time (space) and zero motion, then jumps to maximum time and zero time-of-motion.

- Without parental charge, there is no time (moving time).
- Without resistance, there is no distance. (There are only positions in space.)
- The motion of energy (or matter) is the shortage of time (space).
- The shortage of time draws in parent energy.
- Density is the shortage of space per volume of parent energy.

Review

A system is comprised of a positive circling field of energy that sets the time-of-revolution and the charge (density) of the system. This system also has a lateral negative field of energy that moves through the center of the circling positive field. This lateral negative energy expands as it comes out of the center of the system. This expansion causes it to collect parent energy as it moves out into the parent field. Then as it circles back to complete its circuit, the collected energy condenses and forms charge on the positive field. All forms of motion are displacements of energy. The energy that is moved combines with the energy that it moves into. (Matter is composed of concentrated groups of motional systems.)

Motion (negative lateral energy) and density (positive circling energy) in systems work together to build a continuous exchange of motion-for-density and density-for-motion. When this action is not accomplished in a stationary position because of a shortage of parent density, the systems (matter) reach, or transport, toward the extra needed motion to compensate for the shortage of parent density. (As explained in the Deficient Parent Energy section.)

The Sun, which resides in the more negative galaxy-energy-field, has this shortage of parent density, which causes it to reach for more motion (expand). The circling solar field energy, which causes a boundary gap around the Sun, causes the Sun to expand even farther. This boundary gap around the Sun in the area of the Solar ecliptic plain is what pulls the tail from comets as they pass around the Sun. This condition of matter is called transporting, and it is caused when a boundary gap is formed on one side of a system or on one side of matter. The boundary gap around the Sun is in the galaxy field. These atoms of the comet are now in a system gap as well as a boundary gap. This causes a much greater pull. This pull is not toward the Sun, but it is toward the boundary gap. This pull not only holds the Sun in its place, it is also an expanding pull that converts the gases of the comet back into first or second-field energy.

Basic Energy Reality

There are two kinds of charge, but the same energy. These kinds of charge are positive energy and negative energy. Positive energy is the circling condensing energy field that drives negative energy. Negative energy is the lateral expanding field of energy that draws from parent energy.

Energy freely gives up space (volume) and time (time-of-revolution) of its energy for density and charge. Time and space are the same thing. To shorten time is to shorten space. This action is what happens between the galaxy and the Solar System, the Solar-System and the atom, and also between the atom and its particles. These actions could be called the growth of matter. As these systems get smaller, their speed gets slower. From the Solar System down to the uranium atom, the speed drops from 97,680 feet per second to 1.3 feet per second. When a system (atom) is broken up, its energy is released back into its parent-field (solar field). This causes the energy to expand by the square of its motion until it reaches the expansion and speed of 97,680 feet per second.

This can be accomplished with a slow decay, a slow burn, or an explosion. After reaching this speed, it will stop moving on its own and continue moving with its parent energy.

Thrust

Before energy can expand, it has to be condensed. There are two kinds of thrusts. The one thrust that we are all familiar with is the pushing thrust that happens when energy is being expended. There is another kind of thrust comprised of energy that does the exact opposite. This opposite is a pulling thrust that occurs when energy is time-shortened (made denser). To make parent energy denser is to separate a portion of the parent energy and move the separated portion. This extra motion (which is extra density) separates the two fields, making one more positive than the other. This causes an attraction between them that pulls them toward each other. This is the pulling thrust that creates matter (systems). It is also the pull that lifts airplanes.

Released Energy

As energy expands, it encloses more space. More space means less density. Less density means faster motion. This process of expanding and speeding up continues until the energy reaches the motion and density of its parent field. The solar field energy has a charge of 97,680 feet per second of motion in this diameter of the Solar System. This is where energy from Earth goes as it is released.

Collecting Energy

We cannot collect energy in the sense of reducing its volume. Neither can we destroy or do away with it. With enough knowledge, we can convert it to do almost anything. Nature converts second-field density to third-field density by shortening its time. It is this process that we can use to convert second-field density to fourth-field density (same as the photon). This will give us a whole system gap of positive charge and negative pull to move us at the speed of light. This can be done in two ways. One way is to place one whole system inside of the field of a larger system. Then place another system inside of the field of the smaller system (a system in a system in a system). The other way to build this charge or a greater charge is to place half-systems on a common ring of energy that circulates through the center of a string of fixed positive rings or disks of energy. This will be a type of man-made atom or energy charger. This is what nature does with energy—it just changes energy into matter. All we have been doing is releasing it. When the lateral negative energy of a system is accelerated, it expands and collects more parent energy. When the positive energy of a system is accelerated, it shrinks and becomes denser.

It is these two counterbalancing actions of a system that hold charge and motion together. The reason that a positive disk shrinks when its circular motion accelerates is because the extra motion shortens its time-of-revolution. The diameter of the disk gets smaller to match its new time. This causes the energy to condense into a smaller volume (more charge).

Positive and Negative Energy

The *source*, the *beginning*, and the *reason* for these powers of action are to hold matter together. These powers of positive and negative energy pull parent energy into a formation and keep it restrained until it is released. The following is an explanation on how matter is put back together after it has been released:

Negative energy and positive energy are the same energy. The only difference between the two is that negative energy is time lengthened, while positive energy is time shortened. The negative energy arcs farther out into space before it completes its *time* of cycle. This gives it more time per cycle. (It has less energy per volume of space, which makes it negative.) The positive energy moves in a smaller circle (a shorter *time* of cycle). This disk of energy condenses and builds positive charge. The negative lateral field of energy wraps around, and through, the center of this positive disk. (This is the left-handed law of current and magnetism.) The negative field of energy expands and becomes more negative as the positive disk condenses and becomes more positive. It is these two actions that keep all systems in balance and in motion. To cause this system to move from its stationary position is to speed up the motion of its positive field (disk), making it denser. This causes the negative field to be more negative by the same amount and to pull harder on its parent-field. Another way to get the same action is to make the parent solar field less dense. This causes the positive field of the atom to be relatively denser by the same amount. This is what happens when a photon of light with its fourth-field density leaves an atom and moves into the Solar System and its second-field density.

The secret to free energy is that you pay for it once. Once, to start the system, and then it is free as long as you let the system run.

Space Drive

Space drive is already a fact of nature. It is demonstrated in the action of kinetic energy that keeps objects moving through space. Kinetic energy is caused by small systems (atoms) in matter that aligns with the movement of matter and continues pulling. A photon of light is a half-system that is formed from excess charge that is released from the atom to keep the atom in balance. The only difference between systems is their size and relationship to one another. *Space drive* should be called *space pull* because parent energy never pushes. Only systems that have time boundaries push against one another. The secret of space pull is that all of matter is in constant pull on the parent energy field. This pull is normally from all directions. Kinetic energy occurs when this pull is directed in one direction by moving the matter. Extra charge on either field—the positive sibling-field or the negative parent-field—causes this pull toward each other. When the negative parent-field is made less dense, it is given a greater negative charge. Less density on the negative parent-field is relatively the same as more density on the positive sibling-field. Making the negative parent field *less* dense is accomplished when a photon of light leaves the charge of the atom field, then moves into the *less* dense solar field, then moves into the *less* dense galaxy field, and then moves into the *less* dense universe or (deep space).

Each time a photon of light moves into a more expanded system, its speed is squared as it moves through space. This means that by the time it leaves the galaxy, its speed has been squared at least twice. This is at least 345,960,000 miles per second. Conceivably an instantaneous transfer to the next energy field. This shows that the *density* of the *parent systems* is what prevents absolute motion or absolute expansion.

When photons of light move into a smaller denser system, their speed is reduced by the square root.

Growth of Matter

All matter within an energy system is in constant growth and expansion. One of the reasons for this belief is that the planets, as they extend out from the center of the Solar System, are less dense and are gradually turning into gasses. There are two reasons for this loss of density. One reason is that the motion of the circling solar-field energy moves slower the farther out it extends from the center of the solar field. This means less density (charge). The other reason for this expansion is that the electrons in these planets cannot pull enough energy from the partially depleted solar field to compensate for the lack of charge. This lack causes the electrons to reach for more space (the atoms expand). This is the same action that causes hot lava. This constant growth and demise of matter affects everything inside our Solar System (including all measuring standards). Some things grow faster than others. The human body shows this faster growth (aging effects) more than most matter. I think that as the Solar System grows, and because of changes in density of this portion of our Solar System, the human body responds to this growth the same way as our planet responds.

It seems that the spirit (or law of physical existence) can only hold matter together for a limited time of growth and expansion until it has to start all over again. This is especially true for living plants, animals, and humans.

Most people have noticed that when a person is sick or their body is recovering from some kind of trauma, their pain is greater at night than it is during the daytime. Here is one explanation:

At midnight, a person is 7000 miles farther out in space than they are at midday. The reason for this is because of the diameter and the rotation of the Earth. The parent solar energy is less dense as we move out from the center of the solar field.

I have not studied the method that is used for carbon-dating, but I believe that these different dates reflect the gradual loss of density of our portion of the Solar System. If we assume that the above statements are true, it is then possible that the dinosaurs were not as large as we thought. Their bones have been growing all this time.

Space Travel

It is my belief that man can travel through deep space, even to other galaxies, but we are not ready yet. Everything within the Solar System grew up within the system. Everything that it takes to support life came out of the Earth or off of the Earth. This also means that these products—food, water, and air—are all comprised of small energy systems that when digested break up into separate systems. Then as these separate systems lose their spin and charge, their energy is released into the Solar System. This is a constant cycle that changes energy to matter, and matter back into energy.

The substance of physical life is composed of energy. It is therefore necessary for man to develop an understanding of the densities of energy fields and how to apply them to the body to replace the habit of eating food before going farther into space. When this knowledge has been acquired, he will probably also know how to make water out of energy. The universe is full of energy. It just needs to be converted into usable forms. The Earth knows how to do these things. Man is still learning. There is no need or reason to explore our Solar System until we have this knowledge and these capabilities. With the knowledge that is described in this manuscript, we could travel to another galaxy before we could travel to Mars with a rocket. Trips in our Solar System would be weekend excursions.

My Energy Field—An Update

I have been looking for at *least* a partial answer to the physical problem of the *growth and expansion of the atoms of our body.* I have updated my energy system (half-system) and made it easier to use (see the front cover). Also, the drawing at the end of this section shows the direction of its energies and the location of its fields. The lines of arrows are to show the movement and direction of the positive energy as well as the movement and direction of the negative energy. This sketch is shown as a separate but completed system in the similar way that our Solar System might look. This is why it is drawn out as a flat system that is full of motion. If this was a third field system (an atom), then the negative energy would be called magnetism, and it would be in lines. Some of these lines would extend out and loop through other atoms of like charge and pull them in close.

The drawing shows the negative energy condensing at the bottom where the energy comes together and expanding at the top where the energy collects parent energy. These two energy fields that circulate around and through the tube are the same energy fields that hold a smoke ring together. As this negative energy wraps around and passes up through this tube from bottom to top, it emulates the left-handed law of current and magnetism. The negative energy drives the positive circling energy, and the positive circling energy drives the negative lateral energy.

As the negative lateral energy of a system reaches out into space, it expands and collects the less-dense parent energy. This causes it to have more volume of negative energy. When this collected volume of negative energy is returned and wrapped around the positive field of the sibling system, it causes the positive field to condense and form more charge. This positive field pulls in on the energy that is passing up through the center of the system. This causes a vacuum or low density area in the center of the system.

Note: This is the same action that happens in the center of our Solar System that causes our Sun to be held in an expanded state. It is this action also that pulls the tails from the comets as they go around the Sun.

This low density area is not a good place for the human body. Neither is the negative area at the top of this system. *Caution*, I must say that this energy system is experimental, and it could be dangerous to a person's health. I do not recommend getting into any of these fields. If individuals do experiment with these fields, they should start with a slow-moving circle of air through the tube and then stay in the field no more than one or two minutes and then wait. After a couple of days, increase it if you think you should.

I use the positive condensing side of the system occasionally when checking for its reaction to a physical problem. I have probably been too cautious to come to any sure conclusions as to its benefits. However, I have an immediate remission of prostate and bladder problems every time I use the field, and I feel great afterward. The treatment lasts for many weeks. It is my belief that the energies of these fields (when properly studied) will be used for healing purposes.

As you can see in the cover picture, I have made a curved bench that allows me to stay under the rim of the system. This way I can stay in the condensing positive field. It is also best to keep the bench to one side away from the center. Make sure that you have been completely covered with the energy by turning over at the half-time before getting out of the field.

Note: The reason I am explaining about the possible dangers to the body is because this system looks and seems so harmless. There is also no immediate reaction of pain or discomfort to warn you. The reaction starts twenty-four to forty-eight hours after you quit using the field. I call this reaction space sickness. In most cases, it will be mild discomfort. This scares me also since I don't know what is happening. This is something that astronauts will have to contend with when and if they visit other star systems. This is also likely to happen during extended visits to our own planets because of the changing density of our solar field as we move in or out from the center of the Solar System. There is another thing to think about. Our medical doctors have no knowledge about space sickness. If you need help, you are on your own.

At the present time, I am only using the condensing bottom side of the system and that not very often. Since these fields are below the nerve boundaries of our body, they only react on the atomic structure of our body. We cannot feel these energies going through our body, so we have to wait and see what happens. The only sensation that I feel is a slight dizzy feeling in my head as soon as I get out of the field, and it goes away in about one minute. In my experiments, I found out that it is best to cover my whole body in the field even if I have to change positions while there. These treatments are only for two minutes, and I have the field moving fairly slow. *Caution*, do not get into the field too many days in a row without checking for reactions. The molecular changes could be compounded. Always wait at least three days before continuing your experiments.

This present system is lifted by a winch and cable with an attached metal frame. Its driving force is a fan that circulates air molecules through the tube.

In this paper is an explanation on how parent energy (solar field energy) is carried by air molecules (gases). Gases have no connecting nuclei currents, and parent energy cannot pass through them. That is why they are carried by the parent energy, and the parent energy is carried by them. It is for this reason that weather systems and energy systems are always together.

"Arrows show Lateral-energy, and direction of pos. Field."

Condensing Side

Expansion Side

Three Views

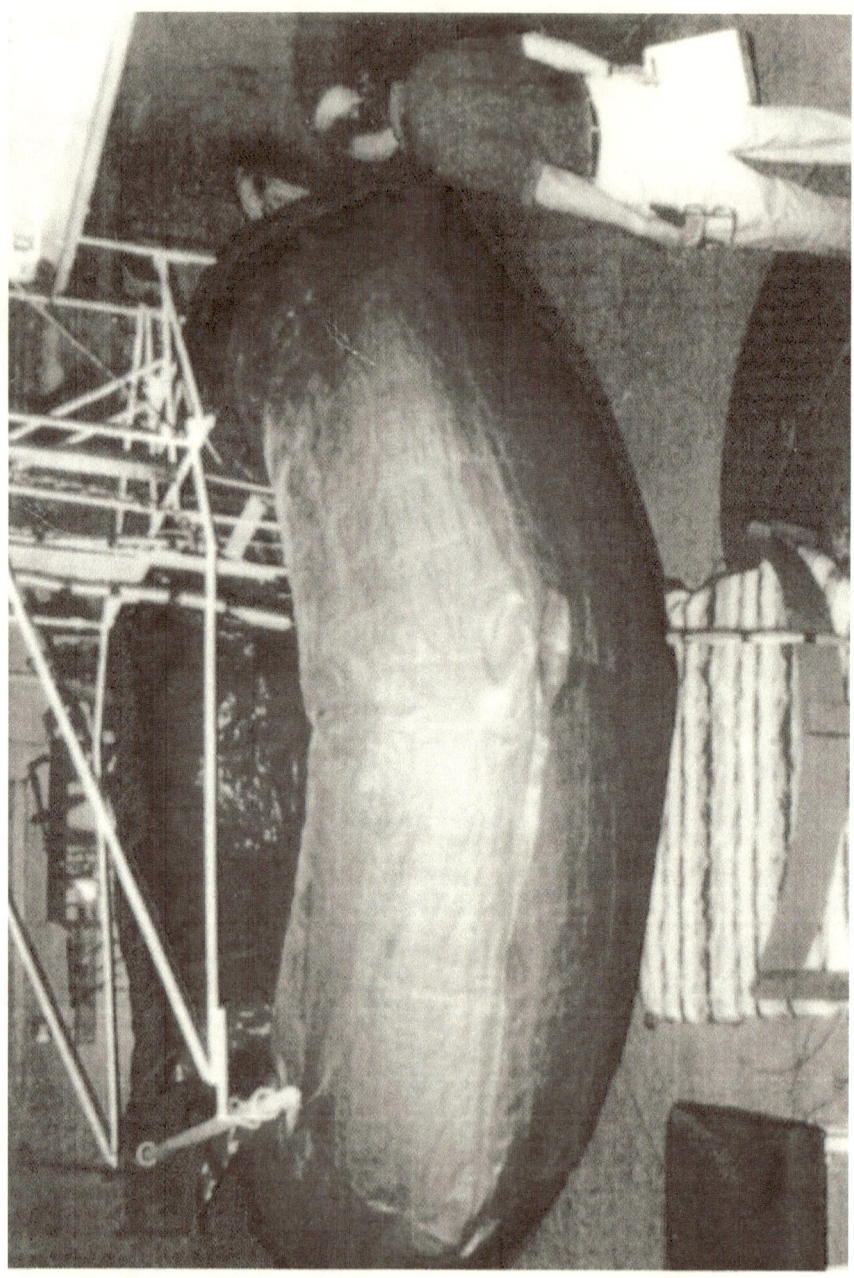

Energy-Field : (System Maker)

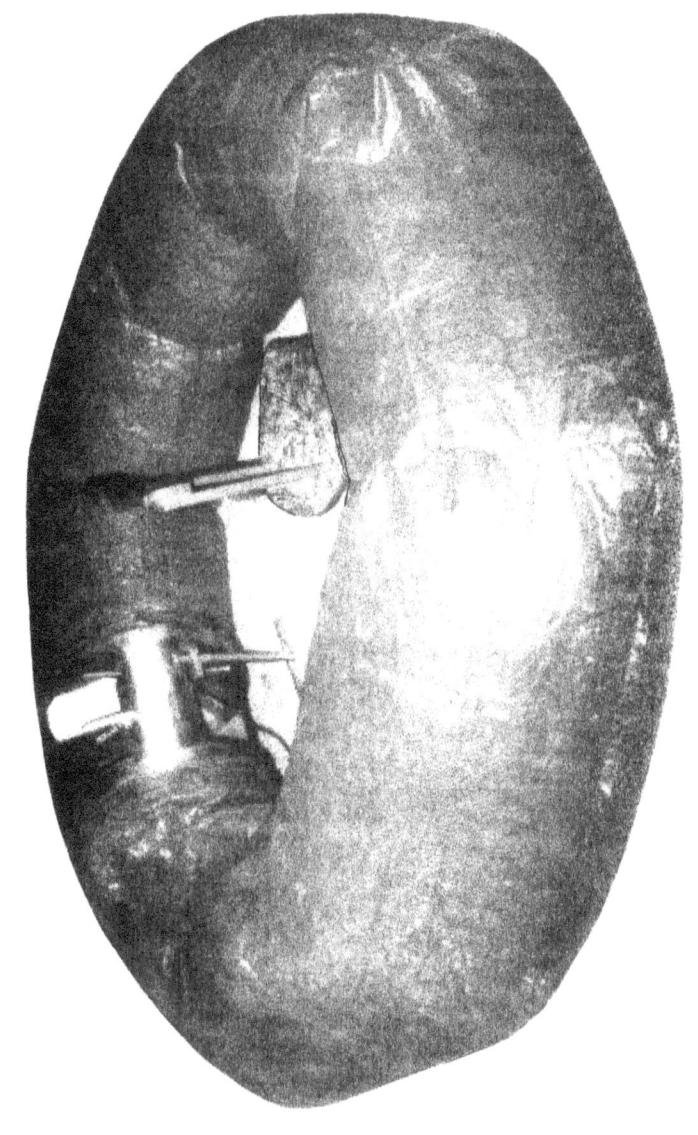

Update on Photons of Light

I read an article about some speculation concerning the movement of photons of light. It seems that the question was "Do photons of light move through conductors faster than they do through the Solar System?" I say the answer is yes. And the reason is because the electrons are pulling on the parent energy that is inside of the conductor, making this portion of the parent energy *less dense*, which causes it to be more *negative* than the parent energy that is outside of the conductor. A photon of light is composed of fourth-field energy and is very dense, very positive, in relation to all fields of negative (expanded) parent energy. The greater the difference of density between a photon and the field it is passing through means the greater the speed of its passing. This is true whether it is inside or outside of a conductor.

Alternating Current System

The reason that alternating current passes through a conductor easily is because the alternating energy is changing directions through the wire at a high frequency, 60 cycles per second or more. This action causes the magnetic lines of energy that circle around the wire to *expand and speed up* until the current reaches its peak in one direction. Then as the direction of the energy is reversed, it does the exact opposite. It *condenses (shrinks), slows down*, and drops to zero. This cycle repeats itself each time the energy changes direction. When negative third-field energy (magnetism) expands, it speeds up and collects parent (second-field) energy. When this negative energy condenses (shrinks), it slows down and converts the collected parent energy to third-field energy onto the sibling system's positive field. This collapsing negative energy keeps converting the collected parent energy to third-field density each time the energy changes direction. Alternating current replaces its energy all along the line each time it changes directions. This does away with resistance.

The way a natural system in nature collects energy is the following: The negative lateral field of an energy system expands and speeds up as it reaches out into space. Expansion and motion of negative lateral energy collects parent energy. This collected parent energy is retimed to third-field energy as it is being wrapped onto the third field circuit and retimed from the motion of the solar field to the motion in the atom.

System Conversions

- The conversion of smaller systems into the larger system: when fuel is burned, the energy is released back into the parent field.
- The conversion to smaller systems from the larger system: when magnetic lines (negative third-field energy) are wrapped around a wire that has its ends connected, it causes the motion of the negative lateral energy (magnetism) to draw in parent energy and form a field of positive energy through the wire that drives electrons around the circuit.

To picture these smaller systems, think of this wire as being so short that its ends will barely meet (typical system). Then stretch this wire to the length of the wire that is in a coil. Thousands of small magnetic rings (systems) circling around the wire causes a large volume of positive energy to move into and through the wire. This positive energy drives the electrons through the wire. This is man-made electricity. The parent solar energy (second-field energy) is pulled in by the third field sibling energy. As this magnetism wraps the parent solar field energy into a smaller circuit, it shortens the parent energy's time (time-of-revolution) from 365 days to microseconds. It also changes the collected parent energy's charge from 97,680 feet per second to 312.5 feet per second (see the Energy Potential section).

Another way to convert to a smaller system is the way gravity does it by pulling on the parent field. As this parent energy comes into the earth, it is separated into microscopic lines of energy that cause counterclockwise energy to encircle these lines. These small half-systems latch onto existing atoms of like-charge, forming more atoms. This causes matter to grow. This is a conversion of second-field energy (parent energy) into third-field energy.

Energy moves in two ways. One way is that it *expands* and speeds up. The other way is that it *condenses* and *slows* down. When energy condenses and slows down, it shortens time and space and builds charge (density). When energy expands and speeds up, it lengthens time and space and builds motion.

Time and Space

Space—Untimed energy (not-moving).

Timed-energy—Energy moving in a circle with its ends connected and forming systems or circuits.

Positive charge—More energy, less space and time.

Negative charge—Less energy, more space and time.

There are two ways to shorten the time of the motion of energy (which builds charge). One way is to speed up its motion. The other way is to shorten its distance of travel. This last way is to loop it into a smaller circle (a smaller system). The first way—speeding up its motion—is by moving the energy or the object (matter) through the charge of its less dense parent-field energy. This object is stationary with its parent energy before it is moved, and they both are moving at 97,680 feet per second around the Solar System. This means that their charges are balanced with each other. As this object is separately being moved from its stationary position, it acquires the exact same amount of motive charge as was used to start it moving. It makes no difference in which direction this added motion is applied, it remains at 97,680 feet per second of charge. This extra motion of impressed negative parent energy on the object has now created a more positive internal charge inside of the object. This added positive charge (which makes it denser), has now caused the negative parent energy to be relatively less dense by the same amount. This causes the object to keep pulling at the same added motion through its parent-field's energy. It has rebalanced its charge to match the two motions.

Buoyancy—The Balance between Two or More Forces

When looking up at a cloud, what we see is tons of water floating above our head. To get the right perspective of this phenomena is to take a bucket and fill it with water and then hold it above your head. Then think of the tons of water that are floating around up there. What we are seeing is called buoyancy between two forces. These forces are positive and negative energy. The following is an explanation: A water molecule is composed of hydrogen and oxygen (basic normal elements). The three electrons that orbit through the center of the molecule do not draw enough parent energy to build a strong nucleus third-field current. It is this negative, cohesive current that pulls atoms together and causes matter to be either solid or liquid. Without this nucleus current, matter is in a gas form. This is what happens to water molecules when heat is added to them. Heat is excessive third-field positive energy without enough negative second-field balance.

When heat is applied, in this case the heat from the sun, this extra third-field energy is added to the positive circling energy of the molecule. This causes the molecules to expand, which pushes their nuclei centers farther apart. This action breaks the already weak nucleus current between them, causing the molecules to become gases and expand. They become half-systems. This means that the molecule's nucleus current has been converted over to and added onto the negative lateral-energy of what has now become a half-system. This half-system is now pulling up (climbing) faster than the parent energy is coming down. These half-systems climb up until the cooler air starts taking heat out of them. At this altitude where the warm water and the cooler air meet is where we see the clouds. The three charges of buoyancy are the positive energy of the half-system, the negative energy of the parent field, and the pull of gravity. The height of a cloud is where this buoyancy takes place.

Spiritual Reality (Life)

It is necessary that this subject, life, be included in this manuscript because energy can do nothing by itself except to grow more systems of like charge. This is why each element has its own charge relative to its number of electrons.

Energy makes up the complete substance of physical reality, including the human body and the brain. The spirit, including the mind, is a separate but attached entity that exists in a different dimension. The spiritual mind has a lot of control over the physical energy (called body) by controlling the brain. To think strong positive thoughts with the mind (called faith) and to speak positive words with the mouth will make a mind-to-brain (spirit-to-energy) connection. These two actions of the will (spiritual mind) cause the physical brain to send coded signals to the parts of the body that the mind is thinking on. These coded signals can be helpful or harmful. They can renew the existing cells of the body, or they can create new and different cells that destroy the body. Fear not, but think on things that are positive and of good report. Fear is a reverse prayer that sends harmful coded signals that can harm you or kill you (my words). The ultimate goal or desire of most, if not all, people (whether they know it or not) is to be able to create with their own mind, to be able to think on things with a positive mind and desire, and have what things they think on to come to pass. This happens every day. Things seem to fall in place at the right time. Connections are made. Arrangements are made easy. Then they look back on what has happened with surprise. There is more to this phenomenon that most people have not thought about. I have noticed that what they picture in their mind is exactly what they get, even to the right color or condition of the object.

This happens even after they have made up their minds to accept something that is not quite what they want. We have a tendency to limit ourselves as to what we think our capabilities and circumstances are. The reason for this is because people don't know about their spiritual capabilities that operate on a higher plane. This spiritual factor is the pure essence of man without his physical body. Man's physical body is comprised of third-field energy.

When and if man reaches his full spiritual potential, he will be able to convert energy into anything that he wants or desires (within certain laws and boundaries) by just speaking to it. As I said, think with a positive mind (believing) to receive. There is also another factor that can come into effect. If another person believes the opposite of what you believe concerning the same thing, it cancels out the two desires. There is a saying for this problem: "Don't let your left hand know what your right hand is doing."

People have to be of one mind and of one accord to accomplish great things. When many people fail to receive what they desire for themselves, or what they desire for another person, it is because they picture in their mind the exact opposite of what they are asking. They ask one thing and believe another. Do not take these people into your confidence. Let them be your left hand. They know not what they do. All people correctly use these laws a few times in their lives just by accident. Successful business people, or I should say anyone that is successful at whatever he or she is doing—and if what they are doing comes easy—use these laws in a limited way. People that are having troubles and problems on a constant basis are not using these laws properly or they are living around too many friends that use left-hand thinking.

These laws that I'm writing about are universal laws of creation. These laws nurture all life forms to either progress, hold steady, or regress. Only mankind has the given ability to choose between progression or regression, life or death. This also includes the ability to prechoose (before death) to take on a better body, or bodies, and to live in an expanded environment called heaven. This is up to each person. We have been given free will to choose. As a born-again Christian, I pray that I have in no way misled anyone or taken words out of context to make my point.

The secret of happiness, which means success, is to know that we are spiritual beings and that these bodies are all composed of third-field energy the same as the furniture we sit on. The brain is our personal computer that controls the rest of the body. We must remember that our brain has been genetically programmed to perform all of the natural functions of the body. The functions of the will—spiritual mind—control all of the other functions like walking, talking, thinking, etc. When our brain (computer) has a problem and cannot send the correct messages to the body, then our body starts to break down. The genetically programmed codes begin to give the cells of our body the wrong instructions. The spiritual mind of this person could actually know what is going on, but it cannot make the brain cooperate. For example, if this person has Alzheimer's, the physical brain cannot understand what the spiritual mind is telling it to do. The person with the handicapped brain has lost some software out of his brain. Now the brain has no record of how it got to where it is or where it came from, and probably does not know what the word *home* means.

About the Author

In the fifth grade of school, I started thinking for myself. By the eighth grade, I was beginning to think like everyone else. I did not want this to happen. I knew that with the ability to read and the desire to learn, I could better choose how to train my own mind, how to think about things without the input from men and women from the nineteenth and even the twentieth century. These thinking men and women (for the most part) have been dead for a number of years. We have brilliant trained scientists and technicians today, but they are handicapped by the shortage *of the knowledge of energy*. While thinking about this mystery and wondering why we have never studied about energy or where energy comes from, it caused me to give up on conventional schooling. I have been thinking for myself ever since that decision. The first thing I thought of was that I needed to acquire knowledge. The second thing that came to mind was that the educational system also needs to acquire knowledge, especially when it comes to science and physics. I realized that it is not the schoolteachers' function to acquire new knowledge. Their role is to pass on to the students what the school board passes on to them. Where, if, and how the school board gets new knowledge and if they pass it on, I'm not aware of it. I do understand how hard it would be to make the right decisions on this because technology (through experimentation) has exceeded our knowledge. We don't know why anything works. We just know how it works. While thinking about the answer to this problem, the solution came to me. When we learn all there is to know about energy, we will not have to experiment on anything because we will know how to do whatever it is that we want to do the first and every time.

To have the complete identity of energy means that all of the answers to all of the why's concerning physics and astrophysics must be related and comparable and must be measurable in quantitative relationship.

www.ingramcontent.com/pod-product-compliance
Lightning Source LLC
Chambersburg PA
CBHW021958170526
45157CB00003B/1044